Contents

Sorting things out

Everything is made of some kind of material. Materials can be **solids**, **liquids** or **gases**.

THINK about different materials.

- A drinking glass is a solid – it has a particular shape.
- **Water** is a liquid – it flows into the shape of whatever is holding it.
- Air is a mixture of gases – a gas has no shape and you cannot see it.

Can you think of other solids, liquids and gases?

You will need:

A pencil and a ruler ✔

A piece of paper ✔

An encyclopedia ✔

Which materials are which?

1. Use the pencil and ruler to divide your paper into three equal columns. Name the first column 'solid', the second 'liquid' and the third 'gas'.

2. Look at all the things around you. What would you say each one was – a solid, a liquid or a gas?

3. Write the names of three solids, three liquids and three gases in the three columns. (Look in an encyclopedia to find out which gases make up air.)

4. Underneath each list, write some words that describe what solids, liquids and gases are like. For example, some solids are hard.

Because . . .

Solids are things that you can pick up because they always have their own shape. They can be hard, like wood, bouncy, like rubber, or soft, like wool.

Liquids are runny and have no shape of their own. They flow into the shape of the container they are poured into.

Most gases are invisible, although you can smell some of them. They have no shape and spread out to fill the space they are in. You use the gases in air to blow up a balloon, for example.

What are solids?

Solids are materials that change their shape only when something else forces them to change.

THINK about how you can change the shape of a solid.

- An apple may be cut into slices.
- A piece of wood can be carved into a toy.

How can you change the shape of solids?

You will need:

Some solids (e.g. a piece of cloth, Plasticine, coins, cheese, a ball, a plastic ruler, wooden sticks, paper, margarine, grapes, peas) ✔

Tools to change your solids (e.g. scissors, a plastic knife, a large stone) ✔

A pencil and paper ✔

Investigating Science

How can solids be changed?

Jacqui Bailey

First published in 2005 by Franklin Watts
338 Euston Road
London NW1 3BH

Franklin Watts Australia
Level 17/207 Kent Street
Sydney, NSW 2000

Editor: Jennifer Schofield
Design: Rachel Hamdi/Holly Mann
Picture researcher: Diana Morris
Photography: Andy Crawford, unless otherwise acknowledged

Acknowledgements:
Spectrum Photostock/Photographers Direct: 12. Ray Moller:3, 6b, 9cl.
Watts: 23b, 24t, 29.Ray Moller:3, 6b, 9cl. Watts: 23b, 24t, 29.

With special thanks to our models: Chandler Durbridge,
Charlie Spicer, Ashleigh Munns, Emel Augustine, Keyon Duffus,
Ben Treasure

A CIP catalogue record for this book
is available from the British Library.

ISBN: 978 0 7496 6091 8
Dewey Classification: 530.4

Printed in Malaysia

Franklin Watts is a division of Hachette Children's Books,
an Hachette Livre UK company.

What can solids do?

1. Test your solids to find out how many ways you can change their shape. Try folding, squashing, stretching, bending, cutting, squeezing, and breaking them.

2. Make a list of your solids and the ways in which each one can be changed.

3. Which solids are difficult to change and which are easy? Why do you think this is?

Because . . .

The shape of some solids is easy to change because the solids are soft or bendy. Other solids are difficult to change because the material they are made from is very hard – it cannot easily be cut or broken.

Liquids and solids

Liquids and solids can sometimes behave in similar ways.

THINK about what happens when you move milk from a bottle to a glass.

- The milk can be poured.
- It takes the shape of the container that holds it.

Can a solid behave like a liquid?

You will need:

A glass of water ✔
A large bowl ✔
A jar of syrup ✔
A spoon ✔
Saucers containing solids of different sizes (e.g. tomatoes, flour, pebbles, peas) ✔

Can you pour a solid?

1. Tip the glass of water into the bowl. What happens?

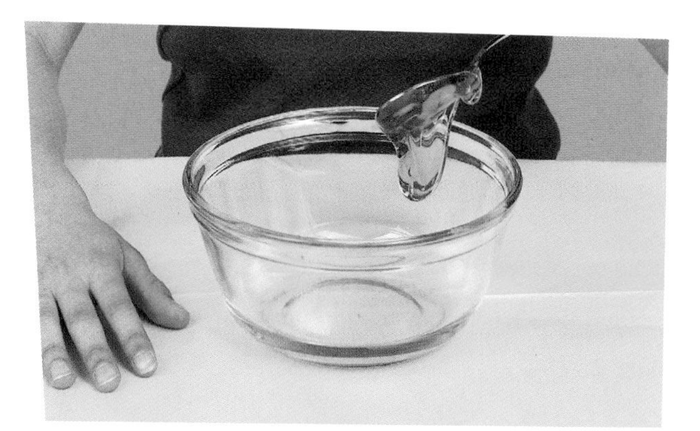

2. Clean and dry the bowl. Tip a spoonful of syrup into it. How is it different from the water?

3. Tip each saucer of solids into the bowl in turn, cleaning the bowl each time.

4. How do the solids compare with the water and syrup?

Because . . .

The tomatoes do not pour because each tomato is a single lump of material. The pebbles, peas and flour look as if they are pouring but only because each small piece is falling at the same time as the other pieces around it. None of the pieces change shape when they hit the bowl. When liquids are poured, even thick liquids like syrup, they spread out to form a level surface within the container.

Changing liquids

Liquids can turn into solids.

THINK about how water **freezes** in the winter.

- When it is very cold, the water in puddles and ponds turns to ice.
- Sometimes it is so cold that drops of water in the sky turn to snow.

What makes a liquid change into a solid?

You will need:

Small, empty yogurt pots ✔
A carton of fruit juice ✔
Aluminium foil ✔
Scissors ✔
Ice-lolly sticks ✔

How can you make a liquid into a solid?

1. Fill the yogurt pots two-thirds full of fruit juice.

2. Cut a square of foil and fold it over the top of each pot like a lid.

3. Carefully use the scissors to make a small hole in the centre of each foil lid. Gently push an ice-lolly stick through the hole into the pot.

4. Leave your pots in the freezer for a few hours – make sure they are upright. What does the fruit juice look like when you take the pots out?

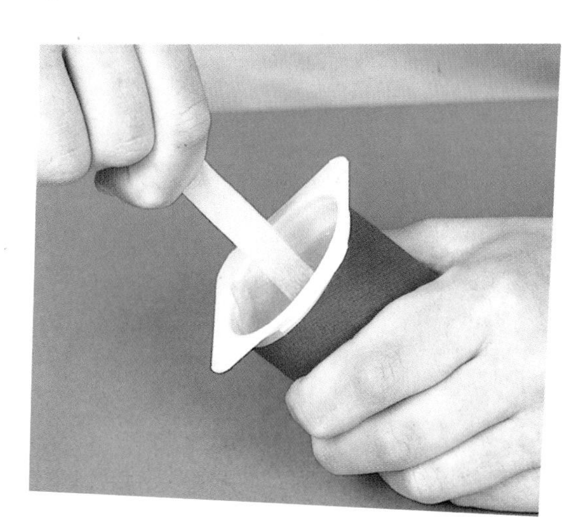

Because . . .

The fruit juice turns into a solid because the freezer makes the juice very cold.

All liquids will turn into solids if they get cold enough. The point at which a liquid becomes solid is called its ***freezing point****. Different liquids have different freezing points.*

Melting solids

Solids can turn into liquids.

THINK about how solids **melt** to become liquids.

- An ice lolly melts when it is taken out of a freezer.
- Chocolate melts when you hold it in your hand.

What makes a solid change into a liquid?

You will need:

A pencil and a ruler ✔

A piece of paper ✔

Some test solids (e.g. an ice cube, a square of butter, a piece of chocolate, a candle) ✔

4 saucers ✔

How can you make a solid into a liquid?

1. Make a chart like the one shown below.

	ice	butter	chocolate	candle
10 mins				
20 mins				
30 mins				
40 mins				

2 Put the ice cube, butter, chocolate and candle on to separate saucers.

3 Leave the saucers in a warm place, such as a sunny windowsill.

4 Check the saucers every 10 minutes for the next 30–40 minutes. Every time you check the saucers, record on your chart how the solids look.

Because . . .

The ice cube melts quickly because it has a low ***melting point*** *– the* ***temperature*** *at which a solid turns into a liquid. Butter and chocolate have higher melting points so they take longer to melt. The candle does not melt because it is not hot enough. Ask an adult to light the candle and watch what happens.*

Mixing solids

Solids can be mixed together.

THINK about food made from a mixture of solids.

- Oats, nuts and raisins are mixed together to make muesli.
- Tomatoes, cucumber and lettuce can be mixed to make salad.

Can mixed solids be separated?

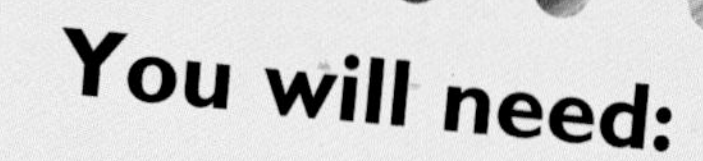

You will need:

2 large bowls ✔

A cupful each of dried pasta, dried beans, oats, raisins, sugar, flour ✔

A colander ✔

A flour sieve ✔

How can you separate solids?

1. Put all the solids in a bowl and mix them together. How does the mixture look? Can you see the separate pieces?

2 Pour the mixture through the colander into the other bowl. What happens?

3 Take away the colander and whatever is in it and pour the second bowl through the sieve into the empty bowl. What happens?

4 Which solids are left in the colander? Which solids are left in the sieve, and which are left in the final bowl?

Because . . .

When you pour the mixture through the colander some of the solids are left behind. This is because they are too big to pass through the holes. When you pour the remaining mixture through the sieve, only the flour and sugar are small enough to pass through the sieve holes. If you had a very fine sieve you could separate the sugar from the flour, too.

Mixing solids and liquids

Solids can behave differently when they are added to a liquid.

THINK about what happens when solids mix with liquids.

- When soil mixes with water it makes mud.
- If you mix powder paint with water it makes liquid paint.

What other changes can liquids make to solids?

You will need:

A large bowl ✔
A cupful of flour ✔
A jug of water and a teaspoon ✔
A fork ✔

How does a liquid change a solid?

1. Pour the flour into the bowl. Can you build the flour into a shape like a cube?

2. Add a few teaspoons of water to the flour and stir it in thoroughly with a fork. What happens to the flour? Can you make a shape with it now?

3. Add more water and keep on stirring. Now what happens?

Because . . .

When they are dry, powders such as flour are difficult to shape because the tiny pieces in them slip and slide around each other. When the flour is damp, all the little pieces stick together and they can be pressed into a shape. When more water is added, the shape collapses. The individual pieces separate from each other and float around in whatever shape the liquid takes.

Dissolving solids

Some solids mix so well with liquids that it looks as if the solid has disappeared. When this happens, we say that the solid has **dissolved**.

THINK about adding sugar to a drink.

- If you stir sugar into a hot drink, the sugar dissolves and seems to disappear. The sugar is **soluble**.

What other solids are soluble?

You will need:

A pencil and ruler ✔
A piece of paper ✔
A jug of water ✔
Some empty glass tumblers ✔
Some test materials (e.g. salt, peas, pasta shapes, soil, powder paint, sugar, instant coffee, sand, flour) ✔
A teaspoon ✔

Which solids dissolve?

1. Use the pencil and ruler to divide your paper into three columns. Head them: 'material', 'guess' and 'result'.

2 Make a list of your test materials in the first column. In the second column, tick the test materials that you think will dissolve.

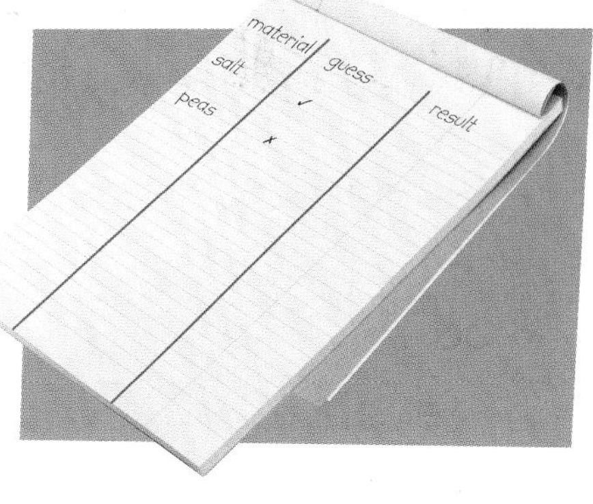

3 Half fill each glass with water. Stir a spoonful of a different test material into each one.

4 What happens to the solid in each glass? In the results column, tick the test materials that did dissolve. How many did you get right?

Because . . .

Some of the solids dissolve in the water. They have broken up into such tiny pieces that you can no longer see them.

Some of the other solids spread through the water and make it look cloudy. These solids have not dissolved, but they are so light they float in the liquid for a while. Eventually they settle on the bottom of the glass, leaving the water above them clear. We call these materials ***insoluble****.*

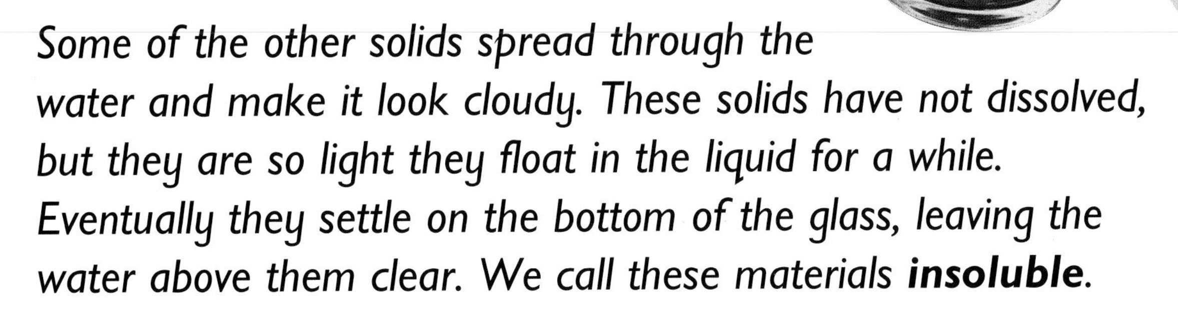

The larger, heavier solids sink straight to the bottom of the glass. They do not mix with the water at all. These solids are also insoluble.

Separating solids and liquids

Some solids are harder to separate from liquid than others.

THINK about what happens when you cook pasta.

- When the pasta is ready, it is poured through a colander to drain away the water before you eat it.

Pasta is insoluble, so it can be separated by straining. How are dissolved solids separated from water?

You will need:

A tablespoon of warm water ✓

A saucer ✓

A teaspoon of salt ✓

How can you separate a dissolved solid?

1. Pour the warm water into the saucer.

2. Add the salt to the water and stir it until the salt dissolves.

3. Leave the saucer near a radiator or somewhere sunny for about 4 hours. Look at it again. What has happened?

Because . . .

The water in the saucer slowly disappears and only small solid bits of salt are left. This is because the water has ***evaporated*** *– the heat from the radiator or sun has turned the water into a gas and it has risen into the air. Dissolved solids can be separated from liquids by evaporation.*

THINK about how washing dries on a sunny day. The sunshine warms the wet clothes and evaporates the water in them, leaving the washing dry.

Heating and cooling

Solids melt into liquids when they are heated. They **solidify** (become solid) when they cool down.

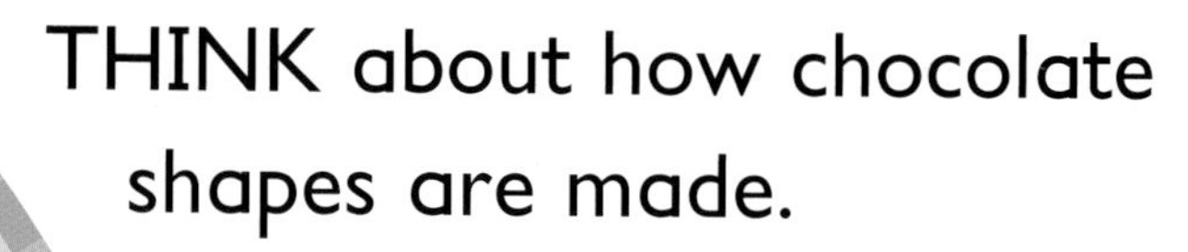

THINK about how chocolate shapes are made.

- Chocolate is heated until it becomes liquid.
- When it is liquid, the chocolate can be poured into a mould.
- When it cools, the chocolate will take the shape of the mould.

Can you change the shape of solid chocolate by heating and cooling it?

You will need:

Greaseproof baking paper ✔
A baking tray ✔
Pastry cutters ✔
Half a bar of chocolate ✔
A glass jug or double boiler ✔
Coloured sweets ✔
A helpful adult ✔

How can heat change the shape of a solid?

1. Cover the baking tray with greaseproof paper. Put the pastry cutters on the tray with the patterned edge face down.

2. Break the chocolate into squares. Ask an adult to melt the squares in a microwave oven or on the stove until the chocolate is runny.

3. Pour the chocolate into the pastry cutters so that it is about half a centimetre deep.

4. Put the baking tray in the fridge until the chocolate has nearly set. Then decorate the chocolate shapes with sweets and put the baking tray back in the fridge.

5. When the chocolate has set, gently push the shapes out of the cutters. What shape is the chocolate now?

Because . . .

When chocolate is heated, it melts into a liquid. Liquid chocolate can be poured into a mould. When the liquid cools, the chocolate becomes solid again. But the chocolate squares do not go back to their old shape. Instead they take the shape of the mould.

Changed for ever

Some materials can melt and solidify over and over again. This is known as a **reversible change**. Others can be changed only once.

THINK about how solids and liquids change.

- Butter is solid at room temperature, melts when it is heated and then cools to a solid again.
- A raw egg is liquid at room temperature, becomes solid when it is heated, but does not become liquid again when it cools.

How does heating change some solids for ever?

You will need:
- An adult to help you ✔
- 150 g soft butter ✔
- 150 g caster sugar ✔
- A mixing bowl ✔
- A wooden spoon ✔
- 1 tablespoon milk ✔
- 1 teaspoon golden syrup ✔
- 1 teaspoon bicarbonate of soda ✔
- 150 g plain flour ✔
- 125 g rolled (porridge) oats ✔
- A greased baking tray ✔

How does heat change solids for ever?

1 Ask an adult to preheat the oven to 150°C (300°F, or Gas Mark 2). Mix the butter and sugar in the bowl until they are creamy and fluffy. Then stir in the milk, syrup and bicarbonate of soda.

2 Mix in the flour and the oats to make a dough.

3 Roll lumps of dough into small balls and put them on the baking tray, spaced well apart.

4 Ask an adult to bake the biscuits for 20–25 minutes or until they are golden brown, then let them cool.

Because . . .

When the mixture is heated in the oven, all the ingredients melt and mix together so well that they cannot be separated again. The mixture has become a new solid. This is called an ***irreversible change****.*

Useful words

Dissolving happens when a solid breaks up into such tiny pieces within a liquid that the pieces can no longer be seen.

Evaporation happens when a liquid heats up and changes into a gas. When puddles dry up in the sunshine, the water has evaporated.

Freezing happens when a liquid is so cold that it changes to a solid.

Freezing point is the temperature at which a liquid becomes a solid. Different liquids have different freezing points. The freezing point of water is zero degrees Celsius (0°C).

Insoluble solids do not dissolve in liquid. Some liquids are insoluble, too. For example, oil does not dissolve in water.

Irreversible change is when a material cannot be changed back to its original form.

Liquids are materials that cannot hold themselves in any particular shape. They flow into the shape of whatever container they are poured into.

Gases

Gases are materials that have no shape. They spread out to fill as much space as they can. Air is made from a mixture of gases.

Melting happens when a solid heats up and changes into a liquid. For example, when a bar of chocolate becomes hot, it melts and becomes liquid chocolate.

Melting point is the temperature at which a solid changes to become a liquid.

Reversible change happens when materials such as water and chocolate melt and solidify over and over again.

Solidify is when a liquid changes into a solid.

Solids are materials that have a definite shape of their own. Their shape does not change except by some kind of force.

Soluble materials are able to dissolve in a liquid so it looks like they have disappeared.

Temperature is a measure of how hot or cold something is. There are different ways to measure temperature, one way is in degrees Celsius (°C).

Water

Water can be found as a liquid, a solid and a gas.

When water is a liquid, it can flow into and take the shape of any container.

If the liquid water reaches a temperature of 0°C, it will freeze and turn to solid ice.

The ice will melt and change back into liquid water when it reaches a temperature of more than 0°C – this is its melting point.

When liquid water becomes extremely hot, it will boil and turn to steam. Steam is a kind of gas. It is very hot, so do not put your hand over boiling water.

Index